Fluid Power
Educational
Series

Electro-hydraulic Proportional Valves

Joji Parambath

Electro-hydraulic Proportional Valves

ISBN: 9798654190819

https://jojibooks.com

Disclaimer of Liability

The contents of this book have been checked for accuracy. Since deviations cannot be precluded entirely, we cannot guarantee full agreement. Only qualified personnel should be allowed to install and work hydraulic equipment. Qualified persons are defined as persons who are authorized to commission, to ground, and tag circuits, equipment, and systems following established safety practices and standards.

Table of Contents

PREFACE

The proportional valves were developed in the mid-1980s, as a low-cost alternative to the servo valves. The proportional valve uses a proportional solenoid to position its spool at the desired position. They are devices for obtaining the infinitely variable flow and pressure controls. They can be used in the open-loop configuration as well as in the closed-loop configuration. Trends in the valve industry today are towards the use of intelligent hydraulics. With this objective in mind, there is widespread development of proportional valves complete with transducers and electronic regulators.

This book explores the technology used in proportional valves. The book also describes the construction of electro-hydraulic proportional valve systems, the details of various types of control elements, and the characteristics of proportional valve systems.

Many other fluid power topics are given in other textbooks under the fluid power educational series by the same author. A list of all the textbooks is given at the end of the book (Page No. 32). Also, please see the details at https://jojibooks.com

Enjoy reading the book.
Your feedback is most welcome.

JOJI Parambath

Chapter 1 | Electro-hydraulic Valves -Introduction

When a discrete type electro-hydraulic directional control valve operated by the conventional on/off solenoids is actuated, its spool is pushed entirely to its maximum travel, usually against the restraining force of a spring. In other words, its spool can only be set in two or three discrete positions.

The precise control of the position, speed, pressure or force is difficult to achieve in systems using the discrete electro-hydraulic valves, due to the 'jerkiness' of motion that these valves experience. Moreover, the remote control of pressure or flow rate is not possible when using these valves. Therefore, the conventional type solenoid valves are primarily used in less sophisticated systems to get the direction control of the flow. More and more electronics are integrated into the traditional hydraulic valves to improve their accuracy and performance.

Electro-hydraulic valves operated by electronic controllers have been developed to overcome the problems that the discrete solenoid valves encounter and to obtain the automated step-less control of pressure and/or flow rate.

Infinitely Variable Hydraulic Valves

In an infinitely variable electro-hydraulic valve, as used in a hydraulic system, the valve output can be infinitely varied with the input signal applied to the valve. In other words, the valve's response needs to be proportional to the input command signal applied to it. The input signal may correspond to the required position of its spool or any other physical variable from the associated system.

The electronically-controlled valves are classified into the following two types: (1) Proportional valves and (2) Servo valves.

Discrete Valves Vs Infinitely Variable Valves

The most common valve configurations used for hydraulic control systems range from the discrete valves controlled by the electromagnets to infinitely variable valves, controlled by the special proportional electromagnets or torque motors. A comparison of these types of valves is most appropriate for broadening the concepts and is given in Table 1.1.

Table 1.1 | Discrete valves Vs infinitely variable valves

Discrete valves	**Infinitely variable valves (Proportional & Servo valves)**
• Mainly used in open-loop control systems	• Mainly used in closed-loop control systems
• Use conventional on/off solenoids	• Use proportional solenoids or torque motors
• The spool moves to its full stroke	• The spool moves in relation to the input signal
• Do not require spool position sensing	• Require spool position (output) sensing
• Provide slow response	• Provide fast response
• Less accurate	• More accurate
• Cheaper	• Expensive

Proportional Valves Vs Servo Valves

The distinction between the proportional valves and the servo valves is inconsistently stated, but, in general, the servo valves provide a higher degree of closed-loop control. A proportional valve can be seen as a conventional directional control electro-hydraulic valve tailored for obtaining proportional characteristics and is gradually acquiring more and more servo valve characteristics.

Though the objective of the servo valves and the proportional valves are similar, the servo valves are more accurate than the proportional valves.

Chapter 2 | Electro-hydraulic Proportional Valve System

An electro-hydraulic proportional valve combines the conventional solenoid valve with a sensor and electronic controller to obtain the precise control of the associated system. With the integration of the electronic controller in the electro-hydraulic valve, it is possible to control the position or speed of the actuator in the system remotely. Further, the proportional valve can easily be interfaced with Programmable Logic Controller (PLC) or Human-Machine Interface (HMI) for easy and user-friendly controls.

The proportional valves are rugged and high-response devices. They always use DC solenoids for the valve actuation. The solenoid coils in the proportional valves are typically operated with 12V/24V DC power supplies.

The proportional valves were initially designed for the open-loop control systems in less sophisticated applications, but, they are also used in the closed-loop control systems depending on the complexity of applications. The following sections briefly explain these two systems.

Open-loop Proportional Valve System

Figure 2.1 shows the block diagram of an open-loop proportional valve system. It is essentially an arrangement of a proportional solenoid valve precisely controlled by an electronic control unit (controller). The required position of the spool can be set by using the potentiometer in the electronic controller.

The proportional valve consists of a spool that is acted upon by the force of the solenoid against the restraining force of the spring. The solenoid develops a magnetizing force in proportion to the actual current flowing through it. The magnetizing force acts on the spool in one direction and against the opposing spring force and places the spool at the position corresponding to the input current.

The key to the basic operation of the proportional valve is the balance established between the magnetic force and the opposing spring force.

It is to be noted that the feedback of the actual position of the spool does not exist in the open-loop proportional valve. Therefore, the accuracy of the open-loop proportional valve system cannot be maintained consistently. The more accurate closed-loop proportional valve systems have to be used in more sophisticated hydraulic systems.

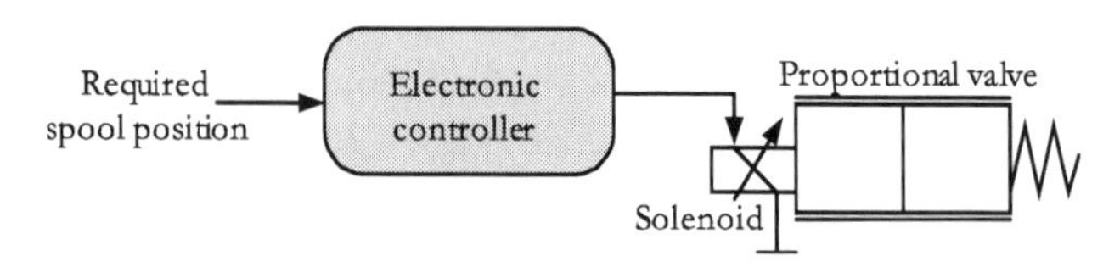

Figure 2.1 | A block diagram of an open-loop proportional valve system

Closed-loop Proportional Valve System

Figure 2.2 shows the block diagram of the closed-loop proportional valve system. It mainly consists of a proportional valve, an electronic control unit (controller), and a transducer (sensor).

The required spool position can be set by using the potentiometer in the electronic control unit. Here, the actual position of the spool can be sensed by the transducer. The output signal from the transducer (or sensor) is fed back to the control unit. A summing amplifier in the electronic control unit calculates the difference between the required spool position (set value) and the actual spool position (actual value). The difference is then fed to the controller. The controller always acts in such a way as to reduce the difference between the set value and the actual value to place the valve spool at the desired position. In this way, more accurate controls can be obtained in the closed-loop proportional valve system. The following section gives the classification of the proportional valves.

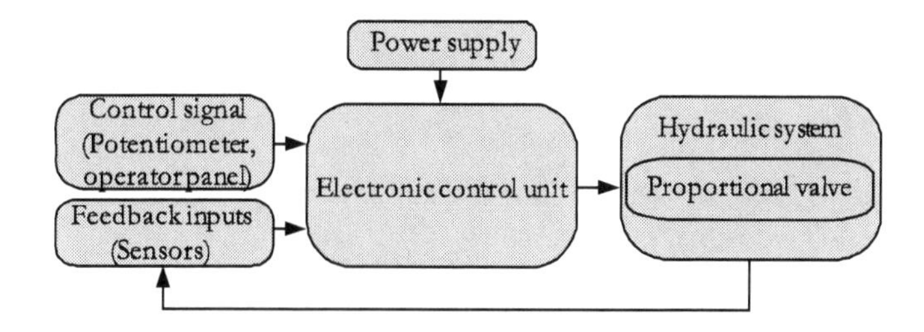

Figure 2.2 | A block diagram of a closed-loop electro-hydraulic proportional valve system

Classification of Proportional Valves

As mentioned earlier, the electro-hydraulic proportional valves provide infinitely variable directional, flow or pressure control functions through the use of the electronic controller units. They can be a combination of these basic types. Accordingly, the proportional valves can be classified as:

- Proportional directional control valve
- Proportional flow control valve
- Proportional relief valve
- Proportional pressure reducing valve
- Proportional directional control and flow control valve
- Proportional flow control and relief valve
- Proportional pressure reducing and relief valve

Chapter 3 | Construction of Proportional Directional Control Valves

The basic proportional directional control valves are low-cost designs without the spool position feedback for the open-loop control applications. They can be the direct-operated and pilot-operated types.

Figure 3.1 details the constructional features of a proportional valve. It consists of a valve body, a spool (or plunger) with restraining springs, and a proportional solenoid with an armature and an electronic control unit. The spool of the proportional valve can be set at any desired axial position within the valve body by varying the solenoid current (command input) as mentioned in an earlier section.

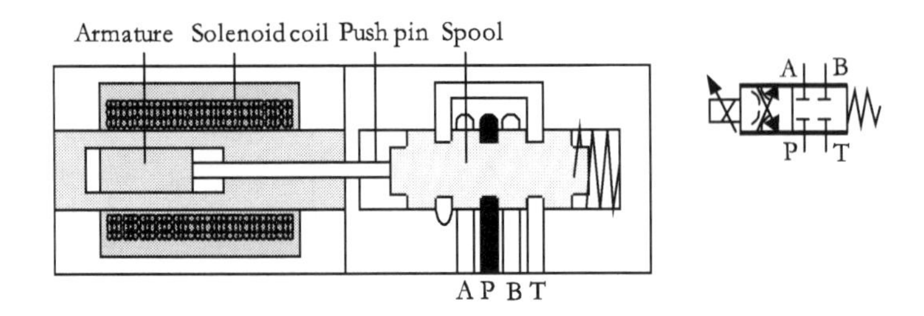

Figure 3.1 | A schematic diagram of the basic electro-hydraulic proportional valve

Figure 3.2 shows the typical metering notches of the proportional valve. It provides a variety of flow profiles in response to the electrical command inputs to its solenoid. The spool and the valve body are designed to open or close the flow path within the valve through the metering notches as the spool is moved through the valve body axially. With the infinitely variable positioning of the spool, it is possible to control the cross-section of the valve orifice available for the fluid flow, either directly or through a pilot stage. The pilot-operated design has a pilot stage and the primary stage. The pilot stage regulates the flow and the pressure acting on the spool of the primary stage.

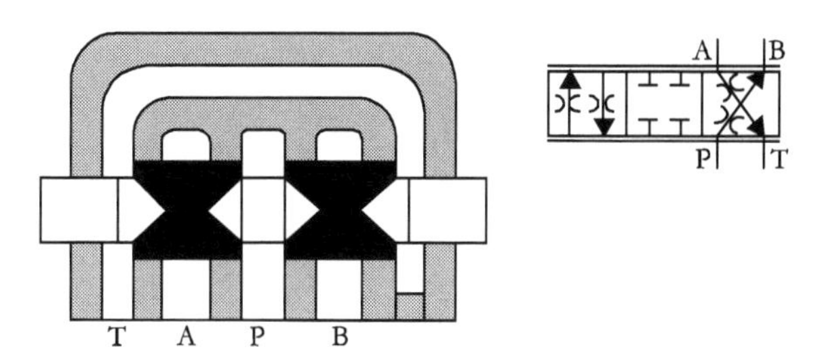

Figure 3.2 | A schematic diagram showing typical metering notches in an electro-hydraulic proportional valve

Deadband

The spool lands overlap the body lands significantly in the centre or null position. It means that the spool of the valve must move a small distance from its centre position before the flow can start when a control signal is applied. This distance, known as the 'deadband', can cause errors and instability in positioning systems. Therefore, to reduce the deadband and to provide the desired offset, a minimum amount of current must be applied to the solenoid coil. This current is usually known as 'null bias'. It is also to be noted that the maximum current must be set to match the maximum working pressure or the maximum flow.

Constructional Features

The proportional control, as specified for the design of the valve shown in Figure 3.1, is possible only in one direction of motion. If the spool is to be shifted to either side of the valve from its centre, proportionally, then two solenoids must be provided, one each on either side of the valve. It may be noted that the present-day proportional valve construction is in modular style with various functional modules, such as valve elements, solenoids, regulating modules, interfacing modules, and power supply modules, manufactured separately and then integrated into a complete proportional valve.

Feedback Mechanisms in Proportional Valves

In earlier sections, we have assumed that the spool position of the proportional valve is determined by the balance between the magnetizing force of the solenoid and the opposing spring force, which act on its spool. However, this arrangement suffers from many critical problems, such as positioning inaccuracy and poor repeatability. Therefore, it is necessary to use the closed-loop control system in the proportional valve to improve its accuracy and repeatability.

The closed-loop control system can be realized by removing the restraining spring in the valve and then adding a positioning sensor/transducer at the end of the valve spool. The controller of the valve gets their position or pressure feedback signal either from the transducer that is mounted on the associated cylinder or valve or from the encoder that is mounted on the axis of the respective rotational system, such as a hydraulic motor.

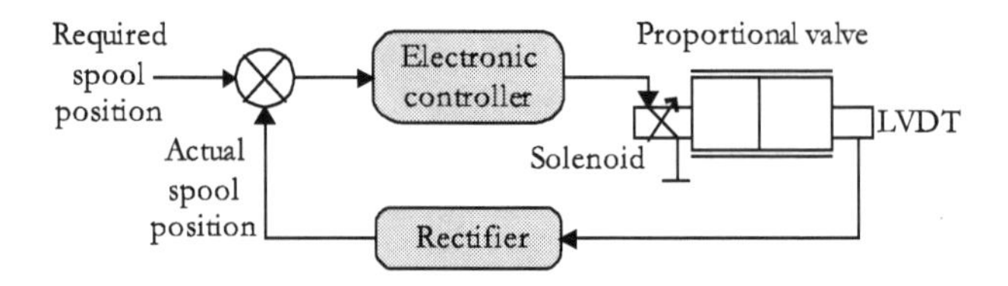

Figure 3.3 | A block diagram of a closed-loop electro-hydraulic proportional valve system

Figure 3.3 shows the block diagram of the closed-loop proportional valve system, consisting of a proportional valve with a transducer (LVDT), an electronic controller, and a rectifier. The spool position of the valve is measured and then converted into an equivalent electrical signal by the transducer. The electrical signal corresponding to the spool position is then fed back into the valve controller through a rectifier, where the controller adjusts accordingly to linearise the flow.

Some examples of the transducer/sensor for measuring the linear movements are Hall Effect spool-positioning sensor, linear variable differential transformer (LVDT), and magnetostrictive displacement transducer (MDT). Remember that a transducer, in general, converts the energy from one form to another form, whereas, a sensor converts the measure of a physical quantity into the corresponding electrical (or pneumatic) signal.

Linear Variable Differential Transformer (LVDT)

Figure 3.4 shows the schematic diagram of the linear variable differential transducer (LVDT). It is an inductive type position sensor used for measuring a linear displacement, such as the spool position of a proportional valve.

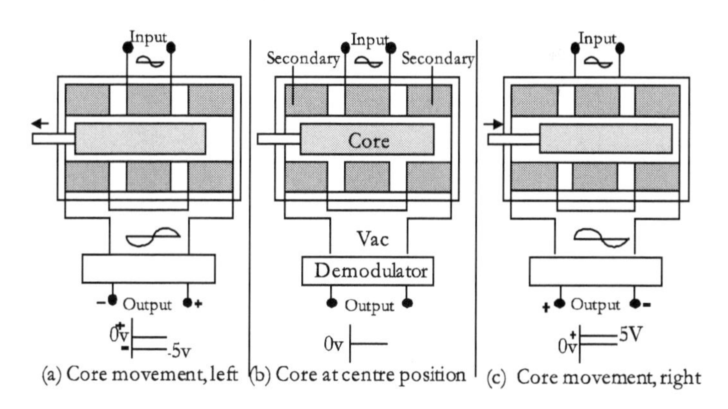

Figure 3.4 | Illustrating the function of an L V D T

It consists of a winding system with a primary coil placed in the middle and two identical symmetrically-spaced secondary coils on either side of the primary winding, surrounding a soft iron core. The coils are wound on a thermally-stable glass-reinforced polymer piece of hollow form, encapsulated against the ingress of moisture, wrapped in a highly permeable magnetic shield. It is then secured in a cylindrical stainless steel housing. A movable ferromagnetic core, which is attached to the spool, slides inside the windings.

The input power to the primary coil is obtained from an electronic oscillator. This oscillator supplies a low-amplitude, high-frequency AC supply (2 – 20 V, 2 – 20 kHz). Voltages are induced in the secondary coils by induction.

The two series-connected secondary windings are 180^0 out of phase. That means; the induced voltages in these windings cancel out each other and produce a zero output voltage, as illustrated in the schematic diagram of Figure 3.4(b) when the core remains in the centre position (null position).

As the core is moved away from the null position, the voltage induced in one of the secondary windings increases and the other secondary winding decreases. These voltages now produce a net output voltage across the series-connected secondary coils. The amount of spool movement determines the magnitude of the output voltage, and the direction of motion of the spool decides the phase of the output voltage.

This fact is illustrated in the schematic diagrams of Figure 3.4(a) and (c). The output voltage can then be fed to a phase-sensitive rectifier (demodulator), which produces a DC signal representing the position of the core.

LVDTs have many advantages owing to their principles of operation, and the type of materials and techniques used in their construction.

Some of the most significant benefits of LVDTs are: (1) friction-free operation, (2) fast, dynamic response, (3) null point repeatability, and (5) long service life.

Next, a brief idea of potentiometers is given before describing the electronic control unit.

Chapter 4 | Control Elements

Potentiometer

A potentiometer is used to give the command signal for setting the required spool position in a proportional valve.

Figure 4.1 shows the two alternative schematic representations of the potentiometer. It consists of a resistor and a wiper placed on the resistor. The position of the wiper determines the voltage (V_o) at the wiper terminal. The voltage must be linearly related to the position of the wiper.

The ideal function of the potentiometer is to provide a command input voltage to the connected amplifier.

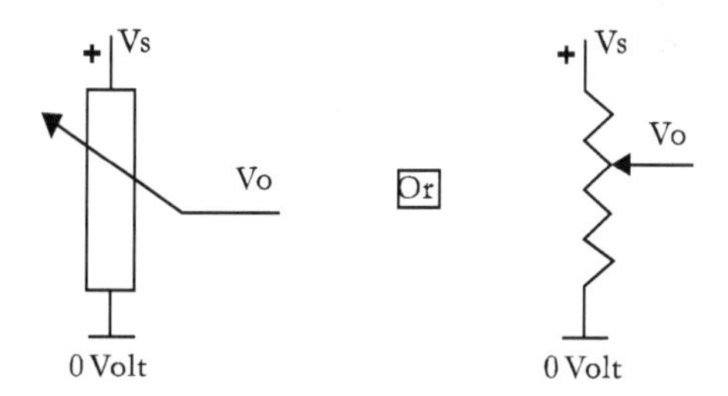

Figure 4.1 | Schematic diagrams of the potentiometer.

However, in practice, a signal voltage at higher level results in the development of a higher voltage drop across the upper part of the potentiometer, and a consequent reduction in the output voltage of the potentiometer. This voltage drop gives rise to a nonlinear relationship between the wiper position and the wiper terminal voltage V_o. The nonlinear relationship in a potentiometer can be reduced by selecting an appropriate potentiometer resistance. As a rule of thumb, the maximum resistance of the potentiometer should be about 10% of the input resistance of the associated amplifier, to restrict the non-linearity of the potentiometer within 2 to 3%.

Joystick

A joystick is a mechanism to translate the movement of a stick into electronic information for the control of machines, such as cranes and robots. The rotational positions of the joystick can be measured using potentiometers, optical or magnetic sensors, and incremental encoders. The measured values correspond to the positions of the joystick and act as the command input signals to the connected amplifier. The most critical parameters of the joystick are its linearity, resolution, size, and cost.

Electronic Control Unit

The electronic control unit in a proportional valve system is the brain of the system.

Figure 4.2 shows the functional schematic of a typical control unit. An essential element in the control unit is an amplifier that converts the low-power input signal into an output signal sufficient to operate a proportional solenoid coil.

The input signal may come from several devices, including a potentiometer or a joystick. The control unit for a closed-loop proportional valve system is designed to receive the spool position feedback from a position sensor.

The electronic control units also include many features, such as ramping (acceleration and deceleration limiting), dither, and pulse width modulation (PWM), and current sensing circuitry to control the associated proportional valves accurately.

The output current in the control unit is usually modified using the ramping, dither, and PWM techniques.

The maximum output current is limited by the setting the potentiometer for the 'Current maximum, I_{max}' to prevent the overdriving of the valve, reduce the effects of temperature change on the solenoid, and protect the valve against electrical short circuits.

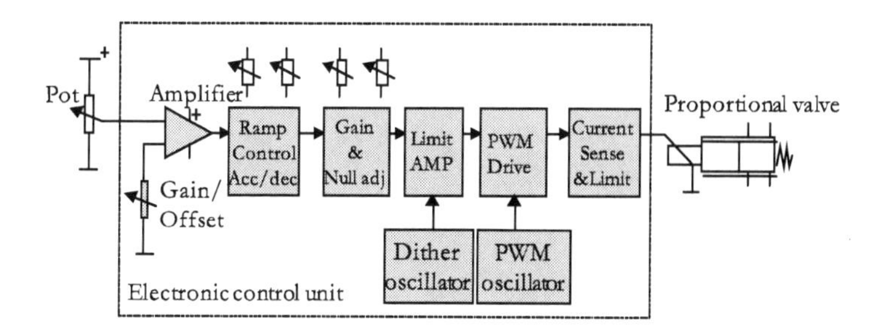

Figure 4.2 | A block diagram of an electronic control unit

The command signals in proportional valves may be bipolar voltages (Ex: ±10V) or unipolar voltages (Ex: 0-10 V) or currents (Ex: 4 -20 mA). In many applications, the current input signal is preferred to the voltage input signal to avoid the effects of large voltage drops across long control lines. Typically, 4 mA represents the zero signal, and 20 mA represents the maximum signal. Any current below 4 mA is ignored to make the system less sensitive to the electrical noise or to provide an indication of the system malfunction. The following section discusses the fundamental principles of electronic amplifiers.

Proportional Amplifier

The proportional amplifier is designed to convert the low-power electrical command signal into a high-power current signal to run the associated proportional solenoid. As the input signal for controlling the proportional solenoid comes from a low power device, such as a potentiometer or a sensor or a transducer, it is necessary to amplify the input signal in order to operate the proportional valve. An electronic amplifier using a transistor is capable of producing a large current flow through its output in response to a small voltage at its input. Figure 4.3 shows the schematic diagram of the amplifier with input and output terminals and a summing junction.

An important parameter of the proportional amplifier is the open-loop gain (A). It is the ratio of its output voltage to its input voltage. Remember that the amplifier inverts the input signal. Therefore, the output voltage can be given by the formula:

Output voltage = -A x Input voltage

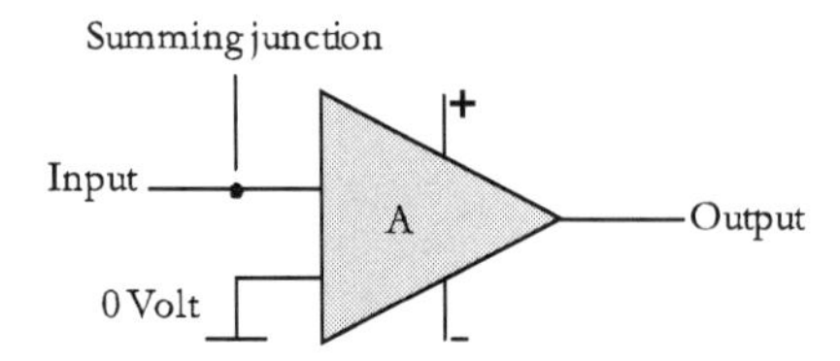

Figure 4.3 | A schematic diagram of a proportional amplifier

The following bulleted lines list the essential characteristics required of the proportional amplifier:

- The amplifier's input impedance must be high so that only a small current should enter the amplifier from the input.
- The amplifier's output impedance must be small so that a large current can be drawn from the output, keeping the voltage constant.
- The summing junction is considered to be always at zero potential.

The type of amplifiers used in control systems is typically referred to as operational amplifiers (OP-AMPs). It is a high-gain amplifier whose operational characteristics are determined by the use of external feedback elements, as shown in Figure 4.4.

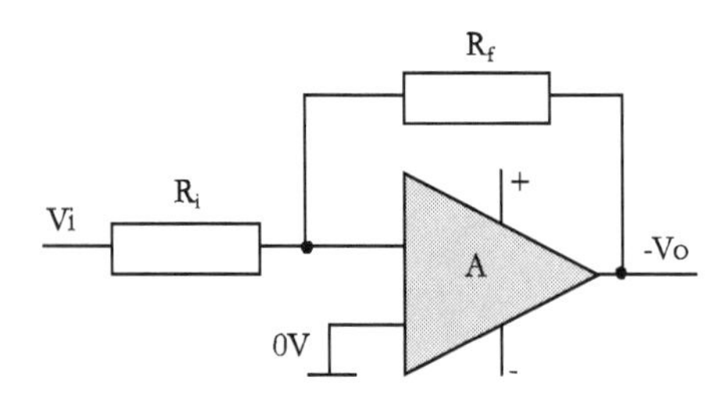

Figure 4.4 | An amplifier with a feedback element

For the amplifier shown in Figure 4.4,

$$V_o = -\frac{R_f}{R_i} x V_i$$

The ratio (R_f/R_i) is the closed-loop gain (A) of the amplifier. As can be seen, the amplifier's gain depends on the value of R_f and R_i. The amplifier must accommodate an AC-operated LVDT, which requires a specialized signal conditioner to supply the high-frequency AC excitation to the LVDT, and a demodulator for demodulating the output of the LVDT to develop a DC voltage in proportion to the spool position.

The amplifier is provided with integrated PID control circuit for the optimum performance of the closed-loop control circuit. See Appendix 1 for the concepts of PID controls.

Additional Features of Electronic Control Unit

In a proportional valve, there are certain additional functions need to be carried out. That is: it may be necessary to control the valve's opening and closing rates.

There is also the need to control the analog type solenoid coil of the valve digitally and overcome the problem of spring hysteresis, stiction, and inertia.

Therefore, the electronic control unit of the valve uses various techniques, such as ramping, pulse width modulation (PWM), and dithering to achieve these additional functions. The following sections explain these techniques.

Ramp Rate Adjustment

The rate of change of the amplifier's output can be controlled through the ramping technique. This method is used to limit its acceleration or deceleration. The ramp rate adjustment ultimately restricts the rate at which the valve opens or closes.

Pulse Width Modulation (PWM)
A primary method for converting the supply voltage to the corresponding output current is by using a rheostat. However, this approach tends to develop a considerable amount of heat and noise.

Therefore, the efficient method employed in the proportional valve for converting the supply voltage to the corresponding output current is by using the pulse width modulation (PWM).

It is a current control method in which a transistor located on the amplifier card of the proportional valve turns the current flowing through the solenoid coil on and off very rapidly to achieve the desired current flow.

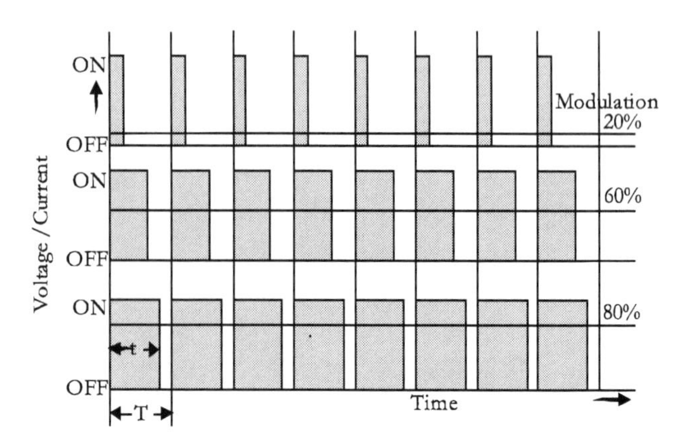

Figure 4.5 | A timing diagram for illustrating the pulse width modulation (PWM) technique

Figure 4.5 shows the timing diagram for illustrating the pulse width modulation (PWM) technique. The displacement of the poppet or spool of the valve can be controlled by varying its 'ON time'. That means if the width of the pulse is 60% of its maximum duration, theoretically the valve shifts enough to deliver a 60% output.

Essential Terms of PWM

The following texts give essential terms of the PWM technique:

(1) PWM period (T) is the time duration of one PWM cycle.

(2) PWM frequency (1/T) is the rate at which the PWM cycles turn on and off in one second, and is given in Hertz (Hz).

(3) The term pulse width (t) is used to indicate the time during which one PWM cycle remains in the "ON" state during one PWM cycle.

(4) The ratio of the ON time of the PWM to its period (i.e., t/T) is regarded as its duty cycle.

Typical Specifications of PWM

The PWM frequency can be low or high. The low-frequency PWM is in the range from 100 to 400 Hz while the high-frequency PWM is in the range from 4000 to 5000 Hz.

Remember, the PWM frequency must be significantly higher than the frequency that the associated hydro-mechanical valve can respond. The valve and hence the actuator can respond only to the average current instead of the instantaneous current.

The high-frequency PWM produces a ripple-free amperage output.

The PWM has become the power amplification method for controlling the proportional solenoids often requiring up to 4 A of current and as much as 30 Watt of power.

Dither Oscillator

A proportional valve operates with small forces derived from the associated solenoid and depends on the small deflections of the valve spool. It is, hence, vulnerable to the problem of spring hysteresis, inertia, and static friction (stiction) between the spool and the body of the valve. The stiction affects the performance of the valve as it causes the valve to ignore small changes in the demanded spool position. The stiction effect is made worse if the valve's spool is held in a fixed position for an extended period, allowing the spool to settle. The dirt in the fluid medium also encourages stiction.

The effects of the hysteresis, stiction, and inertia can be reduced by superimposing a dither signal to the PWM command signal. The dither signal is a low-amplitude, high-frequency sinusoidal (AC) signal (50 – 100 Hz) used for keeping the valve spool in constant motion around its desired position. This signal is too fast for the spool to follow, but the small spool movements are enough to prevent it from staying in a fixed position.

The dither plays a significant role in improving the hysteresis, response, and stability of the valve. Ideally, the oscillation caused by the dither signal does not alter the output of the proportional valve.

The dither in a proportional valve is specified by its frequency (Hz) and peak-to-peak current (mA). The dither amplitude is usually adjustable from 0 to 10% of the rated maximum current of the solenoid. The dither amplitude and the frequency are usually factory set.

Chapter 5 | Characteristics of Proportional Valves

The proportional valve can be configured either as an open-loop system or as a closed-loop system to control the position, velocity, pressure, and/or force of the system. The proportional valve approaches the servo valve characteristics without the associated higher costs.

The solenoid of the proportional valve must produce an output proportional to the solenoid current to generate a predictable response.

There is always a voltage or current offset provided in the controller to ensure a positive shut-off of the valve. Figure 5.1 shows the typical flow characteristic of the proportional valve.

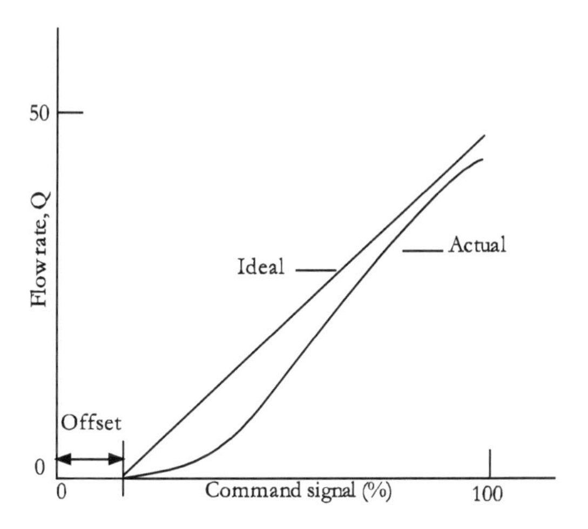

Figure 5.1 | The typical input/output characteristics of an electro-hydraulic proportional valve.

Hysteresis in Proportional Valves

The spool movement in a proportional valve has to overcome some amount of spring force as well as static friction when the input signal is applied. These forces produce an effect known as hysteresis.

Additionally, the solenoid has to overcome the magnetic hysteresis due to the cyclic magnetization and demagnetization of the solenoid core.

The spool movement produced in response to the increasing input signal is not the same as the spool movement created in response to the decreasing input signal.

Hysteresis, as shown in Figure 5.2, is defined as the difference in the spool positions produced by a particular input signal when it is increasing and then decreasing.

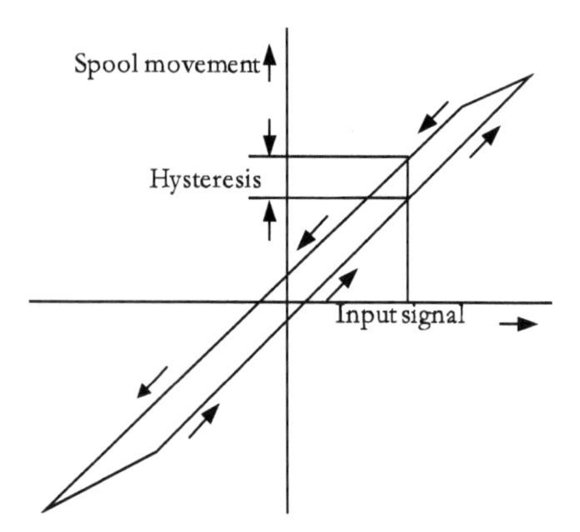

Figure 5.2 | A typical hysteresis curve in an electro-hydraulic proportional valve

Many proportional valves typically use the LVDT spool position sensors with the electronic feedback mechanism. This configuration provides for a tighter spool-position-control-loop and the development of reduced hysteresis.

Chapter 6 | Variants of Proportional Valves

Proportional Flow Control Valves

A proportional flow control valve consists of an electronically adjustable orifice. An external potentiometer can be used to adjust the orifice remotely.

When an electrical signal corresponding to the set value is fed to its electronic amplifier, the controller adjusts the pilot pressure to change the position of the valve's spool. The associated LVDT then produces a feedback signal corresponding to the spool position continuously and sends a signal to the amplifier for a comparison with the set value. The error voltage generated by the comparator drives the controller. The controller, then, adjusts the valve spool to maintain the orifice area for achieving the desired flow.

The proportional flow control valves offer a variety of metering functions with meter-in spools, meter-out spools, combination metering spools, and 2:1 ratio metering spools.

For example, the meter-out spool meters the fluid out of the associated actuator. The 2:1 ratio metering spool provides equal metering in the associated hydraulic cylinder that has a 2:1 piston to piston-rod effective area ratio.

Proportional Pressure Relief Valves

A proportional pressure relief valve limits the pressure in a hydraulic system according to the set value and allows this setting to be adjusted by electronic means. The direct-acting or pilot-operated designs provide a broad range of flow capacities for relief circuits.

Proportional Pressure Reducing Valves

A proportional pressure reducing valve regulates the downstream pressure in a branch circuit according to the set value and allows the setting to be adjusted by electronic means.

Chapter 7 | Applications of Proportional Control Valves

The proportional valves form a large division of the electro-hydraulic controls.

Despite their nonlinear response, the use of proportional valves is an inexpensive way to control the position, velocity, or force on hydraulic equipment for full adjustability and repeatable performance.

Their performance falls in a broad spectrum between the conventional solenoid valves and the servo valves.

They provide precise controls in applications where the hydraulic pressure must be built up steadily, or the system speed must be precisely controlled, or sudden 'starts' and 'stops' must be avoided.

They fit perfectly into the automated control applications. They are becoming more popular for industrial/mobile hydraulic applications.

The proportional valves are suitable for the automotive, marine, and metal fabrication applications.

They are also suited for applications, such as material feeding and edge grinding, press systems, and moulding machines.

8 | Objective Type Questions

1. Mark, the fundamental component used to position the spool of an electro-hydraulic proportional valve:
   a) LVDT
   b) Encoder
   c) Torque motor
   d) Proportional solenoid

2. The distance the spool of an electro-hydraulic proportional valve must move from its null position before the flow can start when a control signal is applied is called:
   a) Threshold
   b) Dither
   c) Offset
   d) Deadband

3. The electro-hydraulic proportional valve system is a combination of:
   a) Solenoid valves, relays and feedback elements
   b) Solenoid valve, electronic controller, and transducer
   c) Torque motor, pilot spool, and main valve
   d) Torque motor, flapper nozzle, and main valve

4. Mark, the feature that is used in electro-hydraulic proportional valves to overcome the problem of stiction
   a) Dither technique
   b) PWM
   c) Ramp control technique
   d) None

5. Mark, the technique in which a transistor in an electronic controller of an electro-hydraulic proportional valve turns the control current 'on' and 'off' very rapidly to achieve the desired current flow:
   a) Dither technique
   b) PWM
   c) Ramp control technique
   d) Current limiting technique

6. Mark, the position sensor that is used in electro-hydraulic proportional valves
   a) LVDT
   b) Flapper-nozzle
   c) Proximity sensor
   d) Limit switch

9 | Review Questions

1. What are infinitely-variable directional-control electro-hydraulic valves?
2. What are the application areas of infinitely variable directional control electro-hydraulic valves?
3. Differentiate the discrete and the infinitely-variable electro-hydraulic valves.
4. Explain the operation of an open-loop electro-hydraulic proportional valve system with the help of a neat sketch.
5. Give a brief note on the impact of electronics on traditional hydraulics.
6. Explain the operation of a closed-loop electro-hydraulic proportional valve system with the help of a neat sketch.
7. Classify the electro-hydraulic proportional valves.
8. Explain the functioning of the solenoid used in an electro-hydraulic proportional valve.
9. What is meant by the term 'deadband' of an electro-hydraulic proportional valve?
10. Explain the term 'null bias' of an electro-hydraulic proportional valve?

11. Explain the feedback mechanism used in an electro-hydraulic proportional valve.
12. What is a transducer? Give an example.
13. Explain the operation of a closed-loop electro-hydraulic proportional valve system with the help of a neat sketch.
14. List the essential components of an electro-hydraulic proportional valve system.
15. What is an LVDT?
16. Explain the working principle of a position sensing device used in electro-hydraulic proportional valves, with a neat sketch?
17. What are the advantages of LVDTs?
18. Explain the function of the electronic control unit employed in an electro-hydraulic proportional valve.
19. Describe various techniques for modifying the output current of the amplifier used in an electro-hydraulic proportional valve.
20. What are the effects of the static friction of the spool in an electro-hydraulic proportional valve?
21. What is meant by 'dithering' and why is it used in electro-hydraulic proportional valves?
22. Write brief notes on the following electro-hydraulic proportional valve terms: (1) Ramp rate control, (2) Dither, and (3) Pulse width modulation (PWM).
23. Briefly explain the hysteresis concerning the electro-hydraulic proportional valves.
24. Explain the operation of a proportional flow control electro-hydraulic valve.
25. Describe the operation of a proportional pressure control electro-hydraulic valve.
26. Explain the operational characteristics of an electro-hydraulic proportional valve.
27. Briefly explain the applications of electro-hydraulic proportional valves.

Objective type questions - answer key:

1-a, 2-d, 3-b, 4-a, 5-b, 6-a

Appendix 1

PID Control

PID controllers are widely used in closed-loop (feedback) control of industrial processes. A PID controller consists of elements with the following three functions. They are: (1) Proportional (P), (2) Integral (I) and (3) Derivative (D).

At first, a designer of a process/system must know its characteristics to control pressure, flow, level, temperature, etc. He must then decide the type of controller to solve the control task. A PID controller as the primary tool in most of the closed-loop control systems.

A Simple Closed-loop Control System

Figure A1.1 shows the block diagram of a simple closed-loop system to control a machine. The control system consists of a proportional gain amplifier and a summing point. The desired operating point that is to be maintained is called the set point (SP). The operating information from the system is called the process variable (PV) or measured variable.

The process value and the set value is compared (by subtraction) to produce a difference called the error. The error is amplified by a proportional gain factor Kp. The amplifier then attempts to adjust the output to reduce the error. In short, the closed-loop system measures, compare and then adjusts.

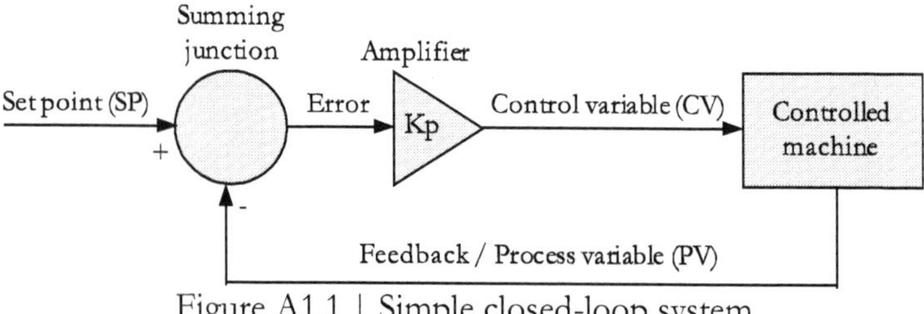

Figure A1.1 | Simple closed-loop system

A simple closed-loop system with a proportional function will rarely work correctly. The proportional gain Kp must be very high to bring the machine's operating point near to the set point. However, when a high gain is used, the system tends to be unstable. As the system tends to approach its set point, the error value also decreases. That means, the system tends to approach the set point at a slower rate. At steady state, the output will level off at some value that is less than the setpoint and will never reach the desired output. The difference between the steady-state operating point and the desired operating point is called offset. (Figure A1.2)

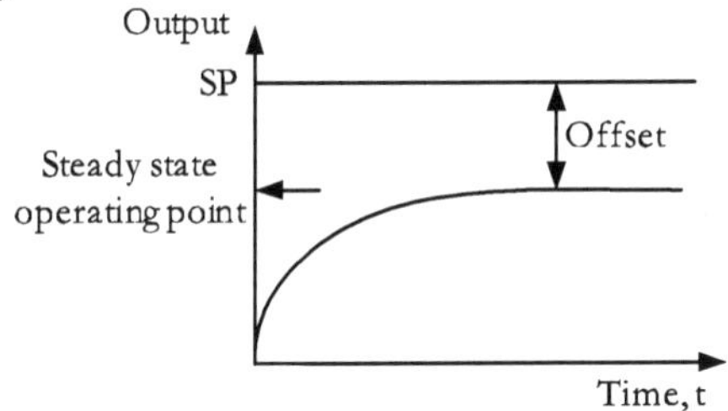

Figure A1.2 | Output response with low proportional gain

It is necessary to add two more functions to the closed-loop system to achieve more desirable levels of performance. They are: (1) Integral function and (2) Derivative function. Each of the functions, namely the proportional, integral, and derivative functions, serves a specific and unique purpose in the system.

Integral Function

The integral function is denoted as $K_i\int$, where K_i is a multiplying constant called integral gain constant. The integral function integrates (sums up) the input (offset/error) over time, and then the result is multiplied by the parameter K_i.

For a small offset, the integral function will accumulate slowly, and its output will increase only gradually. For a large offset, the output of the integral function will change more rapidly to reduce the offset.

In a closed-loop PID control system, the offset can be reduced to near zero by increasing the integral gain constant K_i to some positive value. However, a system with an excessive amount of Ki will overshoot and oscillate.

Derivative Function

The derivative function is denoted by $K_d \frac{d}{dt}$, where K_d is a multiplying factor. The derivative function differentiates the input, and the result is then multiplied by the parameter K_d. For example, when a linear ramp (constant slope) is input to the derivative function, it will output a voltage that is equal to the slope of the ramp. Further, the derivative function will output zero, when a constant DC voltage (slope = 0) is input to it. For more complex waveforms, an approximate derivative can be found by sampling.

The derivative function tends to dampen the acceleration/ deceleration rates. Consequently, the tendency of the system to oscillate is reduced, and the system tends to settle more quickly. In other words, by increasing the derivative constant K_d, the transient response of the system can be improved by eliminating the overshoot and hunting. However, remember that the derivative constant does not affect the offset.

A Typical PID Configuration

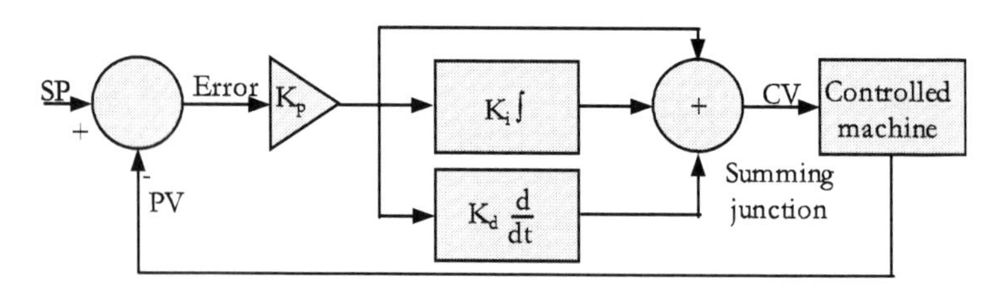

Figure A1.3 | A typical PID configuration

The most commonly used PID configuration for a closed-loop control system is shown in Figure A1.3. The output of the controller can be given as:

$$\text{Output (CV)} = \text{Kp e} + \text{Kp Ki} \int \text{e dt} + \text{Kp Kd} \frac{\text{de}}{\text{dt}}$$

Even though it is possible to get a faster response by further increasing the proportional and derivative gain parameters K_p and K_d, potentially damaging results may occur due to the development of excessive voltage, current, and force transients.

The Behaviour of a PID Controller

A typical behavior of a PID controller for various values of parameters Ki, Ki, and Kd are given in Figure A1.4:

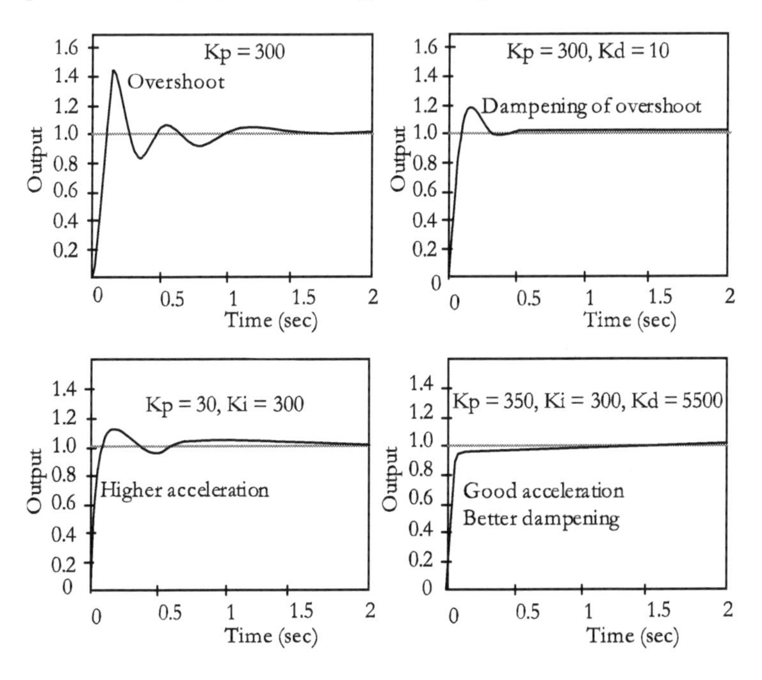

Figure A1.4 | The behaviour of a PID controller

Tuning of PIDs

Tuning of PID controllers involves selecting the best values for K_p, K_i, and K_d. to achieve results that are closer to the desired performance.

The essential characteristics of PID adjustments are as follows:

- Increasing the proportional gain constant (Kp) will result in a faster response and will reduce offset. However, increasing the proportional gain will cause overshoot and hunting.

- Increasing the derivative gain constant (Kd) will reduce overshoot and hunting. However, it will not reduce the offset.

- Increasing the integral gain constant (Ki) will cause the PID to reduce the offset to near zero.

For tuning a PID, the designer must be very familiar with the characteristics of the system. The system must be accurately modelled as far as possible by determining the mechanical and electrical parameters of the system.

Further, the desired response of the system to changes in the setpoint or loading conditions must be evaluated. The designer must also be aware of the theoretical aspects of the PID functions.

10 | References

1. Anthony Esposito, Fluid Power with Applications, 6th Edition, Prentice-Hall of India, 2006
2. Article on: 'PRACTICAL Hydraulic Systems – Operation and Troubleshooting for Engineers & Technicians', by Ravi Doddannavar, Andries Barnard, Elsevier Science & Technology Books
3. Article on: 'Proportional and Servo Valve Technology', Don DeRose, Fluid Power Journal, March/April 2003
4. Bulletin Hy11 PVI017/GB, 'Installation and Setup Manual Electro-hydraulic control for PV series, Design series _ 40, PVplus', Parker Hannifin GmbH, Kaarst
5. Catalogue on DENISON HYDRAULICS Proportional Directional Valves Cetop 07, Series 4DPO3
6. Catalogue on: 'Electro-Hydraulic hybrid system, KAWASAKI ECO SERVO', Kawasaki Precision Machinery Ltd., Japan
7. Document on: 'Electrohydraulic Valves... A Technical Look', MOOG Industrial Controls Division, Moog Inc., NY, USA
8. Document on: 'NEW REGULATOR FOR PROPORTIONAL VALVE-The DSP gives a new heart to electric-hydraulic components', ATOS Spa, Sesto Calende (VA), Italy
9. John R. Hackworth and Frederick D. Hackworth, Jr., 'Programmable Logic Controllers Programming Methods and Applications' Pearson Education
10. Publication on: 'Proportional Directional Valves Cetop 07', Pub. No. 4-EN 3610 –B, DENISON HYDRAULICS

Fluid Power Educational Series Books

1. Pneumatic Systems and Circuits -Basic Level (In the SI Units)
2. Industrial Pneumatics -Basic Level (In the English Units)
3. Pneumatic Systems and Circuits -Advanced Level
4. Electro-Pneumatics and Automation
5. Design of Pneumatic Systems (In the SI Units)
6. Design Concepts in Pneumatic Systems (In the English Units)
7. Maintenance, Troubleshooting, and Safety in Pneumatic Systems
8. Industrial Hydraulic Systems and Circuits -Basic Level (In the SI Units)
9. Industrial Hydraulics -Basic Level (In the English Units)
10. Hydraulic Fluids
11. Hydraulic Filters: Construction, Installation Locations, and Specifications
12. Hydraulic Power Packs (In the SI Units)
13. Power Packs in Hydraulic Systems (In the English Units)
14. Hydraulic Cylinders (In the SI Units)
15. Hydraulic Linear Actuators (In the English Units)
16. Hydraulic Motors (In the SI Units)
17. Hydraulic Rotary Actuators (In the English Units)
18. Hydraulic Accumulators and Circuits (In the SI Units)
19. Accumulators in Hydraulic Systems (In the English Units)
20. Hydraulic Pipes, Tubes, and Hoses (In the SI Units)
21. Pipes, Tubes, and Hoses in Hydraulic Systems (In the English Units)
22. Design of Industrial Hydraulic Systems (In the SI Units)
23. Design Concepts in Industrial Hydraulic Systems (In the English Units)
24. Maintenance, Troubleshooting, and Safety in Hydraulic Systems
25. Hydrostatic Transmissions (HSTs) (In the SI Units)
26. Concepts of Hydrostatic Transmissions (In the English Units)
27. Load Sensing Hydraulic Systems (In the SI Units)
28. Concepts of Load Sensing Hydraulic Systems (In the English Units)
29. Electro-hydraulic Proportional Valves
30. Electro-hydraulic Servo Valves
31. Cartridge Valves
32. Electro-hydraulic Systems and Relay Circuits

For more details, please visit: **htpps://jojibooks.com**

About the Author

Joji Parambath is a trainer in the field of Pneumatics, Hydraulics, and PLC, for over 25 years. During his career, he has trained numerous professionals from the industries as well as faculty members and students of engineering institutions.

At present, he is the key trainer at Fluidsys Training Centre, Bangalore, India, (https://fluidsys.org) which is providing training in the field of Pneumatics and Hydraulics. He has already written two books on Pneumatics and Hydraulics. The publication of the present series of 32 books is intended to restructure and update the existing books.

The author wishes to thank all trainees for their lively interaction and many useful suggestions during the training programmes that prompted the author to write the present series of books. You may send your feedback to joji.p@hotmail.com

10th June 2020

Made in the USA
Middletown, DE
09 October 2023

40271815R00024